YOUR KNOWLEDGE HAS VALUE

- We will publish your bachelor's and master's thesis, essays and papers

- Your own eBook and book - sold worldwide in all relevant shops

- Earn money with each sale

Upload your text at www.GRIN.com and publish for free

Frank Pagram

A Plan for the Cairns CBD Spectacled Flying-Fox Camp. Its Potential for Eco-Tourism and Education

GRIN Publishing

Bibliographic information published by the German National Library:

The German National Library lists this publication in the National Bibliography;
detailed bibliographic data are available on the Internet at http://dnb.dnb.de .

Imprint:

Copyright © 2014 GRIN Verlag, Open Publishing GmbH
Print and binding: Books on Demand GmbH, Norderstedt Germany
ISBN: 978-3-668-01200-4

This book at GRIN:

http://www.grin.com/en/e-book/301740/a-plan-for-the-cairns-cbd-spectacled-flying-
fox-camp-its-potential-for

"Let's welcome our neighbours!"

A plan for the Cairns CBD spectacled flying-fox camp:

its potential for eco-tourism/education.

Contents

This six-week Plan, finalised in October 2014, was prepared for the Environmental Defenders Office of Northern Queensland (EDO NQ). It aims to communicate to EDO NQ and other interested parties the potential for a Communication Tool to reach the key decision-makers listed in the Plan. Its objective is to raise awareness of the important contribution the Cairns Central Business District (CBD) spectacled flying-fox (SFF) (*Pteropus conspicillatus*) camp makes towards Cairns' 'green' image and the associated potential for bat eco-tourism/education (statement of purpose). As will be explained, the emphasis is on Cairns' green image and eco-tourism/education rather than ecological factors.

Desired Outcome

It is envisaged the Plan will be but one step in a much larger and longer plan (*see* Fig. 3), whose ideal outcome is to preserve and enhance the Cairns CBD SFF camp (on Council land and private land owned by the BRCP Oasis Land Pty Ltd, a portion of which is currently vacant/undeveloped) by maintaining the existing large trees on the site(s) with the view to capitalising on the SFF's tourism/educational potential (long-term goal – *see* Fig. 3).

Fig. 3 Flow-chart of stages in the *long-term plan*, clockwise from top. Stages are not in strict order, as different groups/individuals are focusing on different aspects, often simultaneously (including this Plan's author).

AC = Advance Cairns

AUSTROP = Australian Tropical Research Foundation

CAFNEC = Cairns and Far North Environment Centre

Camp = bats' daytime roost for sleeping, resting, protection from predators and weather, grooming, social interaction, information exchange, mating, birthing and caring for young

CBD = Central Business District

CCC = Cairns Chamber of Commerce

CEO = Chief Executive Officer

CRC = Cairns Regional Council

CSIRO = Commonwealth Scientific and Industrial Research Organisation

EDO NQ = Environmental Defenders Office of North Queensland

Flying-fox = fruit bat

GBR = Great Barrier Reef

JCU = James Cook University

PNG = Papua New Guinea

Pteropus conspicillatus = spectacled flying-fox

SFF = spectacled flying-fox

TTNQ = Tourism Tropical North Queensland

WHA = World Heritage area

WT = Wet Tropics

Note that flying-fox 'camp', rather than 'colony', is used in this Plan. This is because 'colony' implies each bat is a somewhat permanent dweller, whereas individual flying-foxes arbitrarily stay at camps for short and long periods, and move among a number of different camps (Ku-ring-gai Bat Conservation Society Inc, 2011). In other words, when bats fly out at night to feed, some return to the same camp, while some go to other camps. Also, the following morning, 'new' bats may arrive from elsewhere.

However, the Cairns CBD camp is different from some other camps in the region in that it is also a nursery, where SFF give birth to pups (September-October) and care for them until they can fly out independently, several months later. It should nevertheless be noted that research and reliable data on Australian flying-foxes are lacking (Westcott, McKeown, Murphy, & Fletcher, 2011).

Current Situation

EDO NQ should have sufficient volunteers, including the writer, to implement the Plan (although it can always use more). While EDO does not have a formal planning hierarchy in which the Plan can be neatly slotted, it may form part of their overall communication strategy on this issue.

EDO NQ has issued press releases, provided legal advice to local residents, and generally raised public awareness of the importance of the Cairns CBD camp. However it has not specifically targeted decision-makers in the way this Plan does.

Acknowledgements

In preparing this Plan, the writer acknowledges the invaluable assistance given, in particular, by Brynn Mathews (EDO NQ), Jennefer Maclean (Tolga Bat Hospital), Angelika Ziehrl (CAFNEC), and Deborah Pergolotti (Frog Safe Cairns). The writer also thanks Dr Hugh Spencer (AUSTROP) for his constructive feedback to the Communication Tool.

The Cairns CBD SFF camp has been used by this species for at least 35 years (Wild, 2012). However, in recent years the SFF population has allegedly increased, and concerns have been raised about noise, smell, bat faeces, and the spread of disease. The camp is located over two CRC Division 5 city blocks (*see* Fig. 4) bounded by Florence St (north-west), Abbott St (north-east) Lake St (south-west) and the Cairns City Library and adjacent buildings (*red arrow*). The Novotel Oasis Resort (*orange arrow*) is within the site. The block on which the resort is situated was once occupied by a primary school, and it is understood some of the (now large) trees were planted many years ago by school children on Arbor Day and when the Oasis was built these trees had to remain.

Land containing the temporary car park to the west of the orange arrow has recently been sold, and an aquarium is planned for this site. The aquarium owners have indicated they plan to retain the existing large trees (an outlying part of the flying-fox camp) and include bat tourism in the complex, with a window looking out to the vegetation and wildlife (Parsons, 2013) .

Fig 4. Cairns CBD SFF camp (original photo: RPS Australia East Pty Ltd, 2013; arrows added by the author).

The vacant land fringed by trees at the opposite (southern) end of the resort from the temporary car park is for sale. To improve the chances of sale, and presumably obtain a better price, the owners, BRCP Oasis Land Pty Ltd, have applied to CRC to remove all trees, as well as some trees around the hotel itself. Both CRC (under threat of legal action from Oasis) and the Federal Government have now approved the application.

Meanwhile, in an attempt to disperse the bats from the site, in 2014 CRC engaged contractors to prune (allegedly not within approval requirements) the large trees adjacent to the library (*labelled red in* Fig. 5) and, in an arrangement with Novotel, on the south-eastern corner of the vacant Oasis land (*labelled yellow*). CRC also removed one of the large trees behind the library (*red 4*), allegedly because it was diseased, but also coincidentally because a new bus station was under construction there. Another tree near the library (*red 5*) was fenced off for CRC removal, due to an independent arborist's report declaring it diseased. However, from September 2014 this work was held up because Federal approval had not been given and flying-foxes were giving birth there. Many of these pups were premature, still-born and deformed, a major cause of concern especially given the camp's trees were 'pruned' six months earlier, possibly causing stress for the newly-pregnant animals (Environmental Defenders Office of Northern Quensland, 2014, 8 October).

Fig. 5. Main Cairns CBD trees occupied by SFF (Wild, 2012).

CRC has not always been antagonistic towards the CBD camp, and even erected an information sign about the SFF outside the City Library a few years ago. However, in 2011 a conservative council was elected, along with a conservative state government, and flying-foxes have also received considerable bad press. Premier Campbell Newman does not appear to be a bat fan, and has pressured councils to take action against them, threatening to step in if they do not do so (CAFNEC, n.d.).

Clearly, if CRC succeed in their attempts to permanently disperse the SFF from the CBD site – an expensive, ongoing action that has no guarantee of success and may result in the bats camping at less-desirable sites – the potential for bat tourism/education is lost. This does not seem to be a concern for CRC Mayor Bob Manning who has been reported as agreeing with TTNQ CEO Alex de Waal who said: "There are, perhaps, some regions in the world that are reduced to the need of focusing on bat tourism", implying Cairns is not one of them (Bateman, 2014).

Moreover, but unlikely to be the major factor in influencing key decision-makers, the SFF is listed as a vulnerable species under the 2009 Australian Environment Protection and Biodiversity Conservation Act (Tait, Perotto-Baldivieso, McKeown, & Westcott, 2014). While its Australian habitat is confined to north-eastern Queensland, it is also found in PNG, the Solomon Islands and Indonesia (*see* Fig 6).

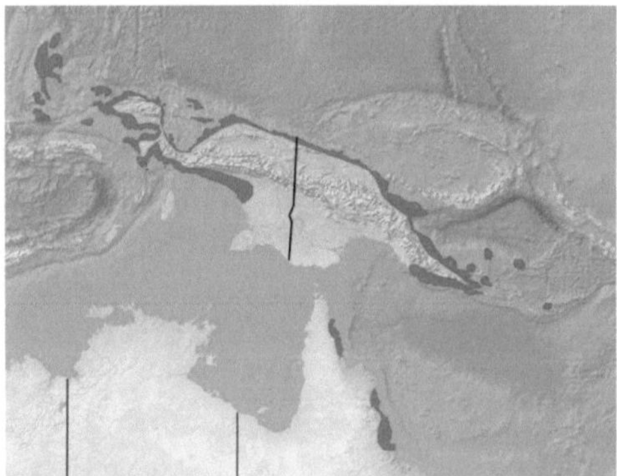

Fig. 6. SFF distribution (Source: https://upload.wikimedia.org/wikipedia/commons/c/c2/Spectacled_Flying-fox.png).

The largest population of the species in Australia is the Wet Tropics rainforests, an often-narrow strip of mountainous and coastal country between Cooktown and Cardwell (*see* Fig. 7). Although it is listed as World Heritage, by no means all of the Wet Tropics has National Park protection.

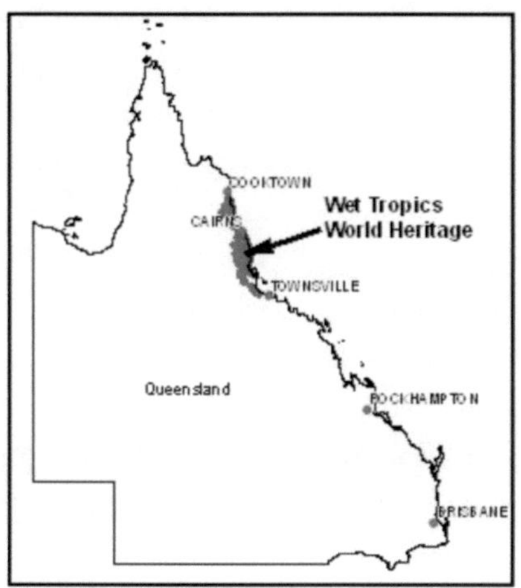

Fig.   7.   Wet   Tropics   World   Heritage   Area   (http://www.wettropics.gov.au/site/user-assets/images/localityqueensland.jpg) .

Largely due to land clearing, the SFF's Australian habitat is diminishing. Flying-foxes are also under threat: from farmers, who at times have used lethal methods to 'cull' bats attacking their orchards; from nets, barbed wire fences, motor vehicles and power lines; and from tick paralysis thought to be caused by the bats consuming non-native tobacco on the Atherton Tablelands south-west of Cairns (Agriculture Resources and Environment Committee, 2012). As a result of these and other threats, the SFF population is decreasing (Westcott et al., 2011).

Because flying-foxes are considered a keystone species – they pollinate, and spread seeds over long distances (not just within forests, but between them and over cleared land), helping maintain rainforest biodiversity – a reduction in the SFF population may have significant implications for the health of the Wet Tropics WHA which many Cairns businesses rely upon.

Yet, despite the efforts of a number of groups and individuals – including local activist Noel Castley-Wright who has made and shared a number of videos, publicised the plight of the Cairns CBD camp and exposed possible breaches of the Environmental Act by CRC and its contractors – the ecological message does not seem to be getting through to key decision-makers; or, if it is, it is regarded as less important than other issues such as the community's health and safety and Novotel's legal threat. Even the *accurate* health and safety message – that the chance of SFF passing Australian bat lyssavirus onto humans is minimal unless the animals are actually handled (in which case a vaccine can be administered) – does not appear to be swaying those in power.

Therefore, this Communication Plan takes a different approach. Rather than focusing on ecological factors, it focuses on economic ones. In particular, this approach was decided upon after a meeting (18 September, 2014) with Angelika Ziehrl from CAFNEC, who advised that the Principal Research Scientist in the CSIRO Ecosystem Sciences' Ecology Program, Dr David Westcott, supported removal of the trees on the Novotel site, and his opinion held weight with the Federal Government who subsequently approved this action. The only recourse, Angelika suggested, was to persuade decision-makers about the potential for bat tourism there.

The Communication Tool emerging from this Plan is a 20-minute talk incorporating a PowerPoint slide-show (20 slides, average 1 minute/slide), followed by 10 minutes' question time. The advantages of this Tool are that it is 'live' (as opposed to a recorded video, although a short video forms part of the presentation) and ideally able to be conducted with most key decision-makers on one, single occasion. This has the potential to capture their collective attention, but will require some organisation and planning. The media should also be invited.

CAFNEC has access to a downstairs meeting room at its inner-suburban base, Cominos House, and a screen and data projector. However, CRC also has similar facilities at its centrally-located chambers in Spence Street and may be willing to make them available for a PowerPoint presentation.

It could be argued there is little point in trying to persuade politicians (some of the key decision-makers) if the public is not behind them, but Cairns politicians have demonstrated public opinion is not the only, or even most important, consideration. Cairns councillors, for example, receive input from a range of people and organisations, including business owners, CCC, AC and TTNQ. Even though polls consistently suggested around 80% of Cairns residents supported retaining the town square, City Place, CRC bisected it with a road in 2014, thus destroying Cairns' only CBD dedicated-pedestrian space. Similarly, a recent Cairns Post survey suggested about 70% of Cairns residents wanted the trees on the Novotel land to stay (Vlasic, 2014), but CRC approved the Oasis application.

Although the Oasis application has been approved by CRC and the Federal Government, tree felling is not permitted until May, 2015, when all SFF, including juveniles born in September/October 2014, should be able to independently fly out at dusk. Under Federal environmental law, no tree may be pruned or felled while any bats are currently observed in it, and if a bat is discovered work must stop immediately.

Thus, there is a half-year window of opportunity to, if not reverse the CRC and Federal Government decision to allow the tree removal, at least put forward a case that it may be preferable to retain the trees and the SFF camp.

Planning Process

The planning process is represented diagrammatically (*see* Fig. 9), with labels then expanded upon (*see* Table of Strategies and Actions, *below*).

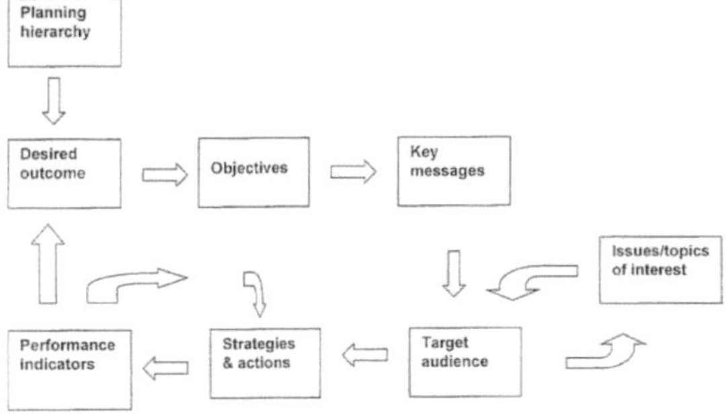

Fig. 9. The planning process (adapted by Queensland Parks and Wildlife Service, 2001, p. 15, from Tamborine Public Contact Plan)

<table>
<tr><td>

Desired Outcome (when combined with future Plans):

To preserve and enhance the Cairns CBD SFF camp by maintaining the existing large trees on the site(s) with the view to capitalising on the tourism/educational potential of the SFF.

</td></tr>
<tr><td>

Objectives (short-term goals):

1. Most key decision-makers accept invitation to the PowerPoint talk.

2. Local media also attend.

3. The majority of key decision-makers undertake to consider the prospect of retaining the Cairns CBD flying-fox camp, including the large trees, rather than continuing efforts to disperse the animals.

</td></tr>
<tr><td>

Key messages:

- SFF are cute, intelligent, and vulnerable.
- They are not dangerous unless handled.
- Flying-foxes are the main pollinators of many Australian native trees. They are also important seed dispersers, necessary for forest biodiversity.
- Many local businesses rely on healthy forests. Forests also provide timber and many other benefits.
- In Australia, the SFF is confined to north-east Queensland, with Cairns being the largest, centrally-located city and thus best placed to capitalise on these animals' tourist potential.
- Cairns' green image is part of the tourist drawcard. It is also why many folk (maybe even flying-foxes!) make this city their home.
- Tourists appreciate the CBD flying-fox camp, often photographing the nightly fly-out.
- Chinese tourists visit Cairns, and organisations such as TTNQ are trying to attract more. For the Chinese, bats symbolise long-lasting good fortune.
- Other parts of the world have capitalised on bat tourism, for example Austin Texas.
- The long-term economic losses of trying to disperse bat colonies may outweigh any perceived, short-term gains.

</td></tr>
</table>

Topic of interest	Target audience	General strategies	Specific actions	Performance indicators
Cairns CBD spectacled flying-fox camp: its potential for eco-tourism and education.	**Key decision-makers:** Cairns-based Federal (Warren Entsch) and State MPs (Gavin King). CRC CEO Peter Tabulo. Cairns Mayor (Bob Manning) and/or Cairns deputy-mayor (Terry James). CRC Division 5 Councillor (Richie Bates) and other Cairns Councillors. CEO's of TTNQ (Alex de Waal), CCC (Deb Hancock) and AC (Mark Matthews). Cairns Novotel Oasis Hotel manager (Peter Richardson). Representation from local Indigenous people would also be sought, as would representation from eco-tour businesses open to the idea of flying-fox tourism/education.	Information dissemination, predominantly via the Communication Tool (i.e., PowerPoint talk).	**Week 1:** Send letter invitations to key decision-makers, requesting RSVPs. **Week 2:** Process responses. Contact respondents who did not accept. **Week 3:** Contact non-respondents. **Week 4:** Process new responses. Contact respondents who did not accept. Alert local media. **Week 5:** Communication Tool: PowerPoint talk, highlighting flying-fox tourism and educational potential. **Week 6:** Plan next action(s).	Majority show-of-hands from key decision-makers attending talk for serious consideration of camp retention.

Monitoring and Evaluation

It could be argued the "majority show-of-hands from key decision-makers" is a simplistic

way of assessing the Plan's success. However, key decision-makers may be reluctant to put

pen to paper endorsing the Plan's statement of purpose (*see* p. 3), whereas a show-of-hands is

more informal and does not commit decision-makers to *necessarily* changing their minds.

Apart from the main Performance Indicator listed above, monitoring can also occur: during implementation of the Plan via the Specific Actions (i.e., the number of 'yes' responses to invitations sent in Weeks 1 and 3); the number of key decision-makers who attend the PowerPoint presentation; and the number/type of questions asked by key decision-makers, the media and others. From this monitoring/evaluation, the next stage in the long-term plan would be sown. This would almost certainly include follow-up with the key decision-makers, unless their response to the presentation was overwhelmingly negative/hostile (considered unlikely). However, in relation to the Oasis site, it should be noted that the owners (located in Sydney and the U.S.A.) are not the hotel manager (located in Cairns); thus any decision made locally may not be supported nationally.

Conclusion

This Plan is a small step towards influencing key decision-makers regarding the Cairns CBD SFF camp. As part of a long-term plan, and with input from other individuals and organisations, the overall aim is to sway these decision-makers to reverse their decision to attempt dispersal of the camp, and instead retain the large trees on the site(s) with the view to capitalising on the SFF's local tourism/educational potential. A suitable outcome could be the establishment of a SFF Interpretative Centre on the (now) vacant Novotel Land, perhaps in a public-private partnership including Indigenous Australians.

References

Agriculture Resources and Environment Committee. (2012). Report No. 14: Land protection legislation (flying-fox control) Amendment Bill 2012. Brisbane, Qld., Australia.

Bateman, D. (2014, 26 April). Tourists love the wildlife says expert but flying fox tourism a batty idea says Mayor, *The Cairns Post [online]*. Retrieved from http://www.cairnspost.com.au/lifestyle/tourists-love-the-wildlife-says-expert-but-flying-fox-tourism-a-batty-idea-says-mayor/story-fnjpuwet-1226896540994

CAFNEC. (2014). Novotel Oasis applies to cut down Spectacled Flying Fox roost trees. Retrieved from http://cafnec.org.au/2014/08/novotel-oasis-applies-remove-cut-spectacled-flying-fox-roost-trees/

CAFNEC. (n.d.). Flying Foxes. Retrieved from http://cafnec.org.au/what-we-do/wildlife-issues/bats/

Department of the Environment. (2013). Pteropus conspicillatus — Spectacled flying-fox. Retrieved from http://www.environment.gov.au/cgi-bin/sprat/public/publicspecies.pl?taxon_id=185

Environmental Defenders Office of Northern Quensland. (2014, 8 October). Cairns CBD spectacled flying fox colony in desperate situation [Facebook status update]. Retrieved from https://www.facebook.com/EnvironmentalDefendersOfficeOfNorthernQueensland/posts/709733675779186

Ku-ring-gai Bat Conservation Society Inc. (2011). Nomadic range and movement. *Sydney bats.* Retrieved from http://www.sydneybats.org.au/flying-foxes/grey-headed-flying-fox/grey-headed-flying-fox-nomadic-range/

Parsons, L. (2013). Cairns Aquarium developers throw support behind flying foxes in the CBD, *The Cairns Post [online]*. Retrieved from http://www.cairnspost.com.au/business/cairns-aquarium-developers-throw-support-behind-flying-foxes-in-the-cbd/story-fnjpusdv-1226748778006

Queensland Parks and Wildlife Service. (2001). Planning workbook: Connecting people with nature through interpretation, extension and community education: The State of Queensland Environmental Protection Agency.

RPS Australia East Pty Ltd. (2013). Cairns Spectacled Flying Fox Colony. Relocation Plan. Retrieved from http://www.environment.gov.au/cgi-bin/epbc/epbc_ap.pl?name=show_document;document_id=53642;proposal_id=6937

Tait, J., Perotto-Baldivieso, H. L., McKeown, A., & Westcott, D. A. (2014). Are flying-foxes coming to town? Urbanisation of the spectacled flying-fox (Pteropus conspicillatus) in Australia. *PLoS One, 9*(10). doi: 10.1371/journal.pone.0109810.

Vlasic, K. (2014, 29 August). Novotel tree clearing stalled after Federal Government asks for more details on flying fox removal, *The Cairns Post [online]*. Retrieved from http://www.cairnspost.com.au/news/cairns/novotel-tree-clearing-stalled-after-federal-government-asks-for-more-details-on-flying-fox-removal/story-fnjpusyw-1227040850522

Westcott, D. A., McKeown, A., Murphy, H. T., & Fletcher, C. S. (2011). A monitoring method for the grey-headed flying fox, Pteropus poliocephalus. Retrieved from http://www.environment.gov.au/biodiversity/threatened/species/pubs/310112-monitoring-methodology.pdf

Wet Tropics Management Authority. (n.d.). World Heritage Area maps. Retrieved from http://www.wettropics.gov.au/wha-maps

Wild, R. (2012). Strategic direction: Flying fox colony Cairns CBD. *Agenda: Cairns Regional Council Ordinary Meeting 31 October.* Retrieved from http://www.cairns.qld.gov.au/__data/assets/pdf_file/0005/63968/31oct12_ordinary_cl14.pdf